VIRUSES

Rebecca Woodbury, Ph.D., M.Ed.

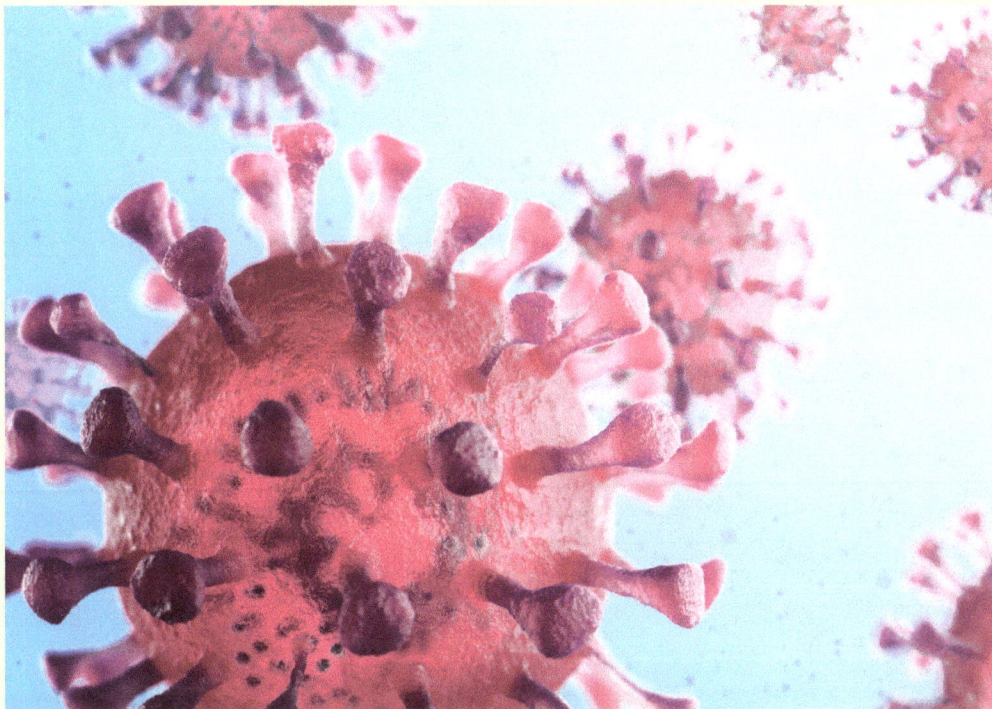

Gravitas Publications Inc.

VIRUSES

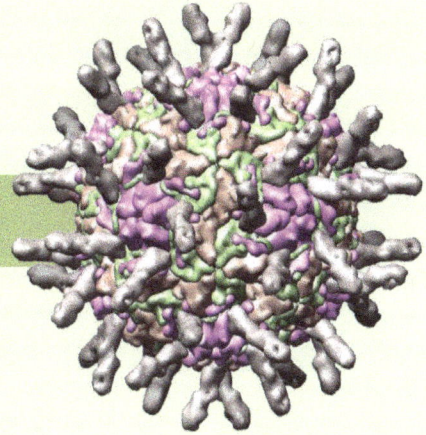

Illustrations: Janet Moneymaker

Copyright © 2025 by Rebecca Woodbury, Ph.D., M.Ed.

Viruses
ISBN 978-1-950415-52-6

Published by Gravitas Publications Inc.
Imprint: Real Science-4-Kids
www.gravitaspublications.com
www.realscience4kids.com

RS4K

Photo credits: Cover & Title Pg: By creativeneko, AdobeStock; Above & P.16. Reference Protein Data Bank structural studies of two rhinovirus serotypes complexed with fragments of their cellular receptor, Kolatkar, P. R., Bella, J., Olson, N. H., Bator, C. M., Baker, T. S., Rossmann, M. G., Journal: (1999) EMBO J. 18: 6249-6259]; P.3. By illustrissima, AdobeStock; P.11. By Vink Fan, AdobeStock; P.17. CDC/Alissa Eckert, MSMI; Dan Higgins, MAMS, Public Domain; P.19. By SeventyFour, AdobeStock

Have you ever been sick
with a runny nose?

Or had a fever?

Or a sore throat?

You may have had a **virus.**

Viruses cause many colds and flu.

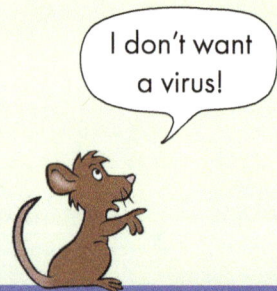

I don't want a virus!

But what are viruses?

In simple terms, **viruses** are like small bags of **molecules** covered with more molecules.

Viruses are too tiny to see with only our eyes.

Wait! What are molecules?

Review: MOLECULES

Molecules are made

when **atoms link** together.

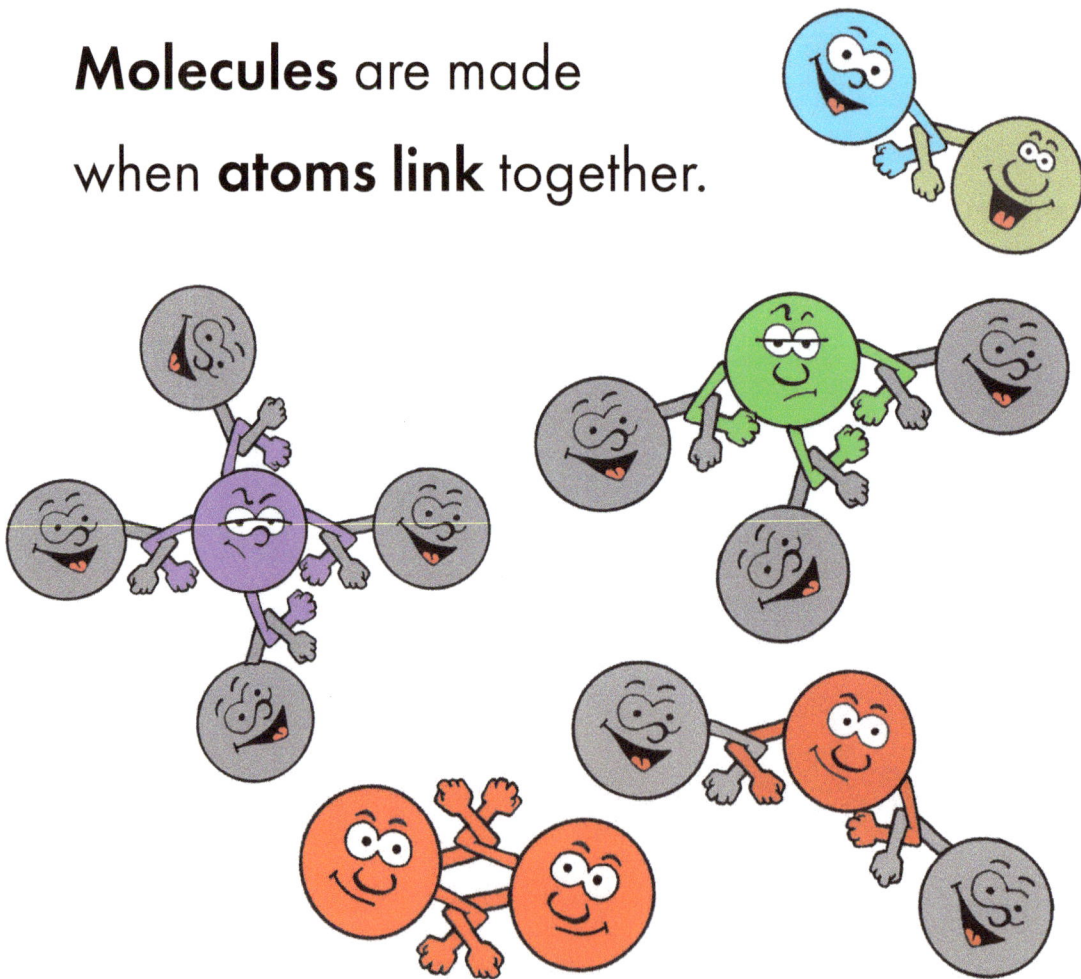

Review: ATOMS

- **Atoms** are tiny building blocks that can link together.

- **Atoms** make everything we see, touch, taste, and smell.

But viruses are not actual cells.

Viruses do not have many of
the parts that cells have.

Viruses cannot do many of the
things cells do.

Uh oh!
Are we missing
any parts?

I think we
have them all.

Review: CELLS

- All living things are made of **cells**.

- **CELLS** are made of **atoms** and **molecules.**

- Each **cell** has many parts that do different jobs.

Because viruses do not have all the parts of actual cells, viruses must live in other creatures to survive.

Washing your hands can help keep you from getting a virus.

We can build a cat trap!

Wait! We don't have all the parts.

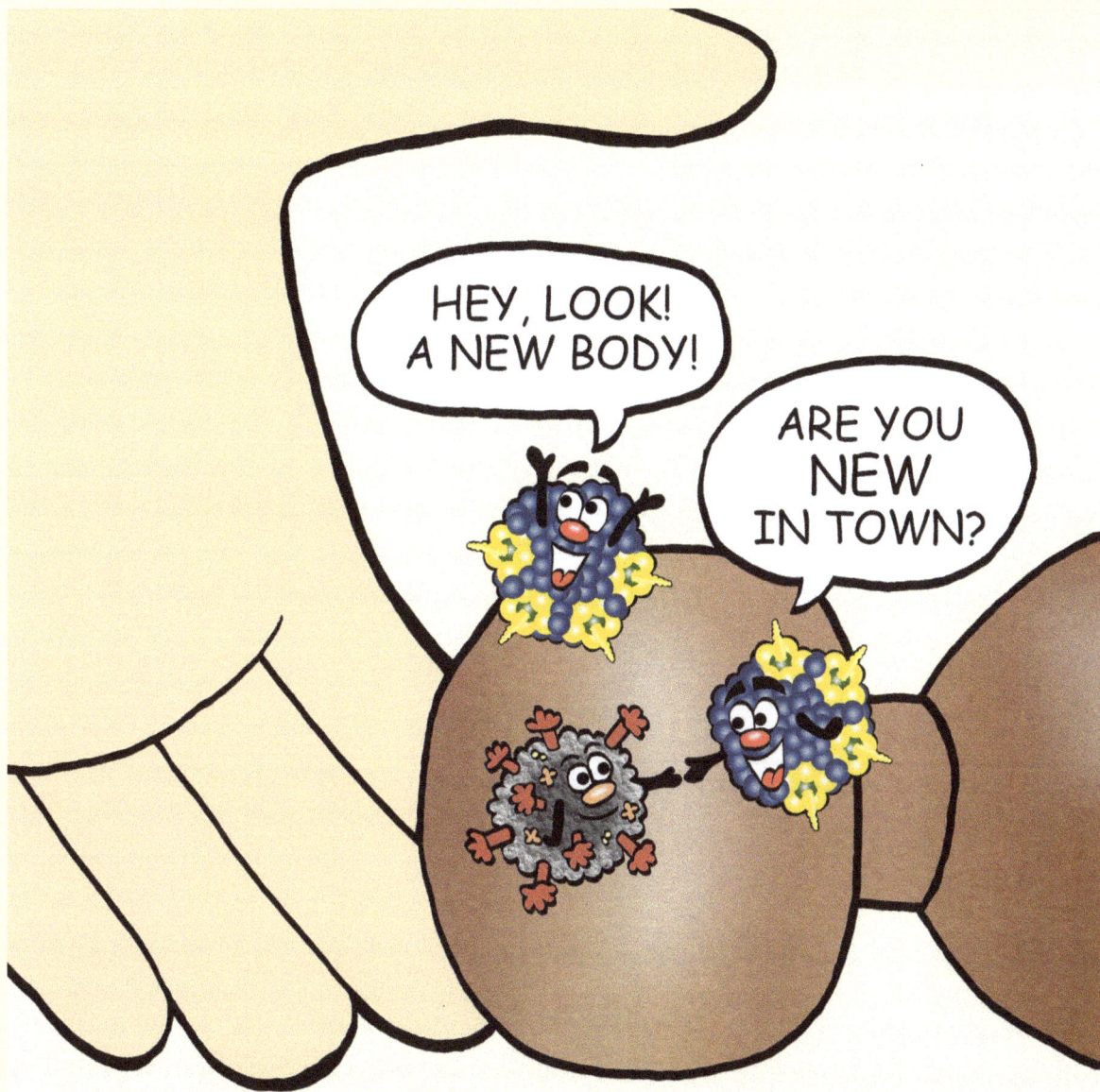

Because viruses must live in other creatures, some scientists think viruses are not alive.

Do viruses need mice to survive?

Maybe!

Viruses come in many shapes.
The common cold virus looks
like a soccer ball with spikes.

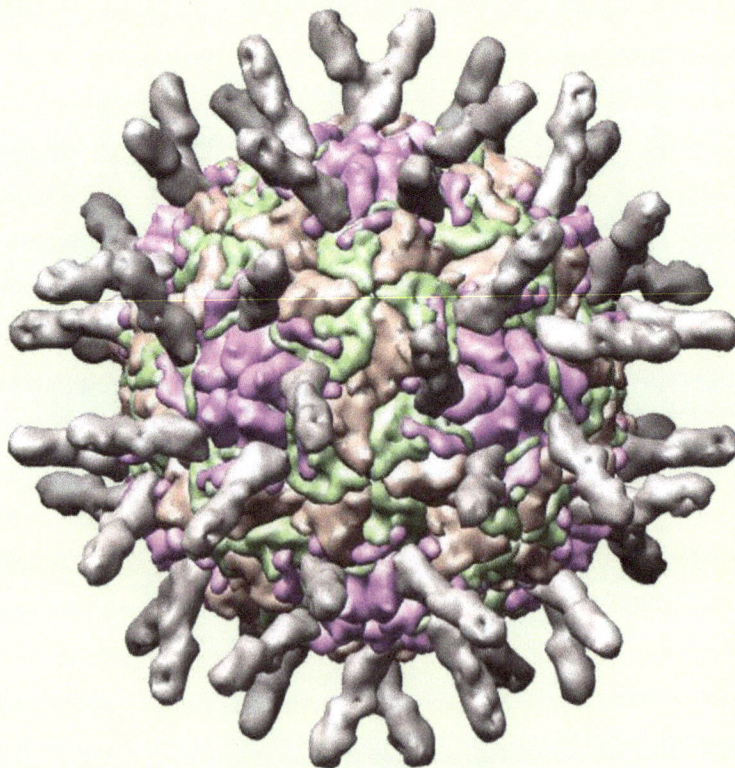

The Ebola virus looks
like a long worm
with clover leaves.

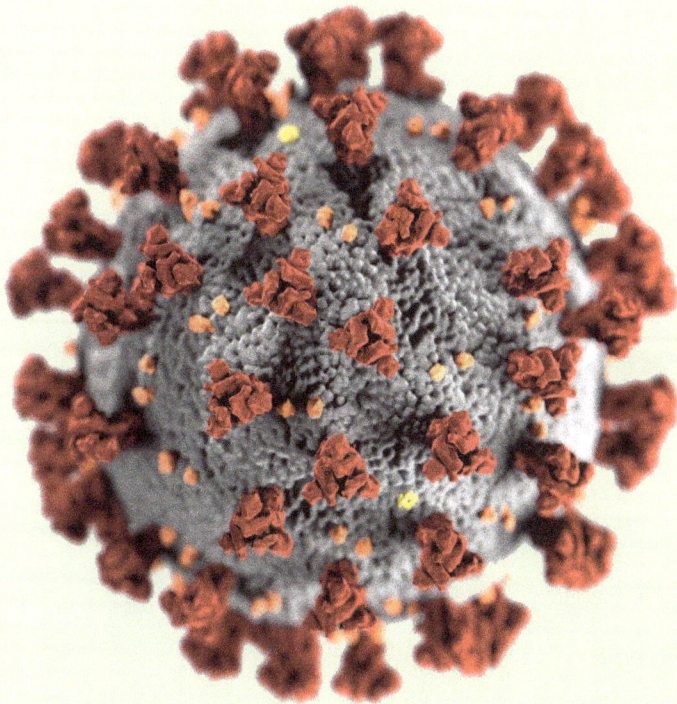

The coronavirus
looks like a fluffy
ball with flowers.

Viruses can cause diseases
in plants and animals.

These plants look like they have a virus.

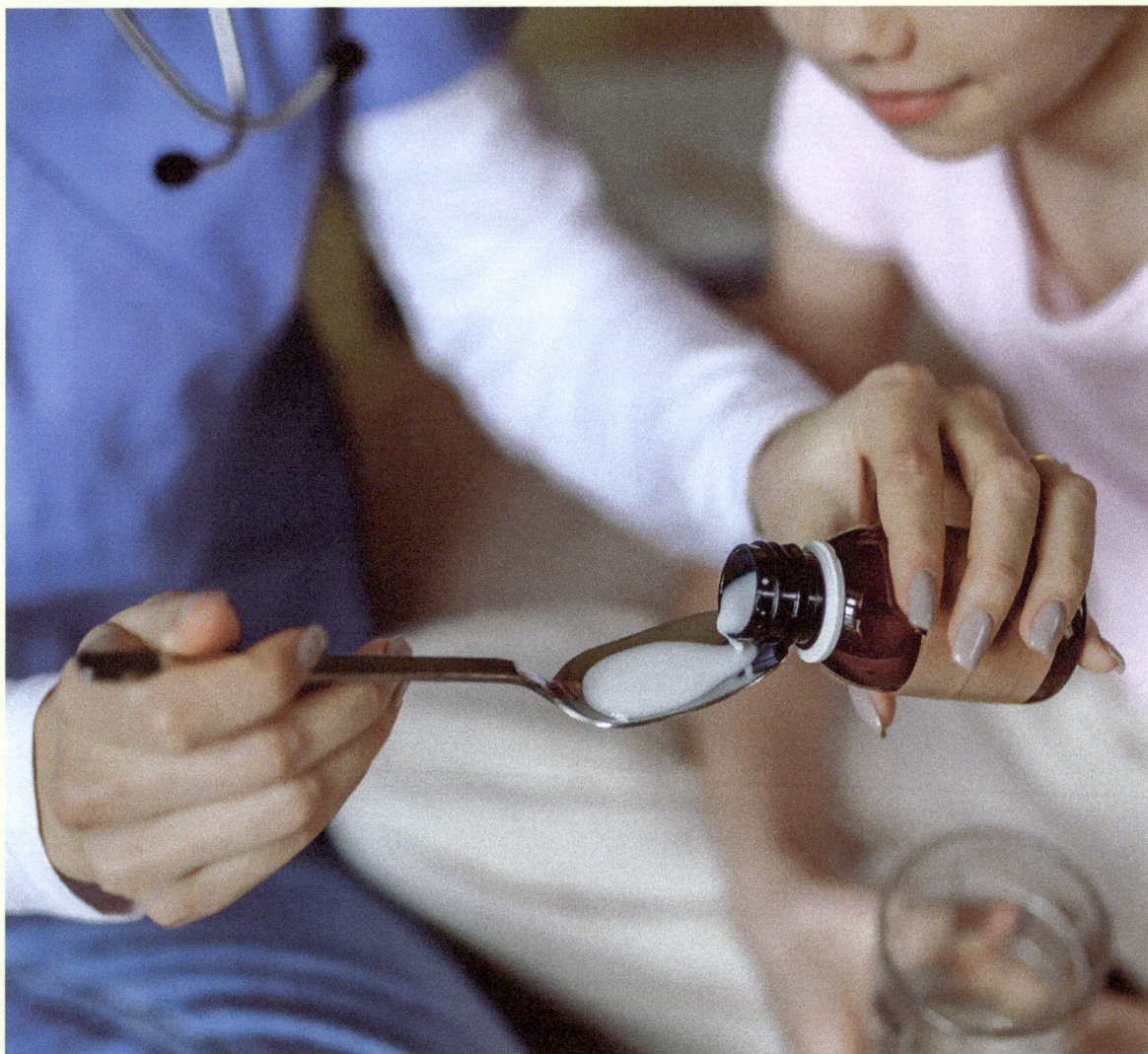

But not all viruses are bad.
Scientists can use some
viruses to cure diseases!

That is good to know!

How to say science words

atom   (AA-tum)

cell   (SEL)

coronavirus   (cuh-ROH-nuh-VIY-ruhs)

Ebola   (ee-BOH-luh)

molecule   (MAH-lih-kyool)

science   (SIY-uhns)

scientist   (SIY-en-tist)

virus   (VIY-ruhs)

www.ingramcontent.com/pod-product-compliance
Lightning Source LLC
Chambersburg PA
CBHW040153200326
41520CB00028B/7588